Dear Parents and Educators,

Welcome to Penguin Young Readers! As parents and educators, you know that each child develops at their own pace—in terms of speech, critical thinking, and, of course, reading. Penguin Young Readers recognizes this fact. As a result, each Penguin Young Readers book is assigned a traditional easy-to-read level (1–4) as well as an F&P Text Level (A–R). Both of these systems will help you choose the right book for your child. Please refer to the back of each book for specific leveling information. Penguin Young Readers features esteemed authors and illustrators, stories about favorite characters, fascinating nonfiction, and more!

Tiny Terrors! The World's Scariest Small Creatures	LEVEL **4**
	F&P TEXT LEVEL **R**

This book is perfect for a **Fluent Reader** who:
- can read the text quickly with minimal effort;
- has good comprehension skills;
- can self-correct (can recognize when something doesn't sound right); and
- can read aloud smoothly and with expression.

Here are some **activities** you can do during and after reading this book:
- Nonfiction: Nonfiction books deal with facts and events that are real. Talk about the elements of nonfiction. Then, on a separate sheet of paper, write down the facts you learned about the following animals: blue dragon sea slug, Irukandji jellyfish, Indian red scorpion, vampire bat, and killer bee.
- Descriptive Words: A descriptive word is one that points out a specific characteristic of someone or something. The author of this book uses a lot of descriptive words to show just how small and scary some of the creatures she writes about really are. For example, the Indian red scorpion's size is described as "itty-bitty" and a vampire bat's teeth are "sharp" and "pointy." Reread the book, pointing out any descriptive words you see.

Remember, sharing the love of reading with a child is the best gift you can give!

For Tanya Anderson, my former boss and forever friend: thanks for the title, inspiration, and encouragement!—GLC

PENGUIN YOUNG READERS
An imprint of Penguin Random House LLC, New York

First published in the United States of America by Penguin Young Readers,
an imprint of Penguin Random House LLC, New York, 2022

Photo credits: used throughout: (photo frame) Tolga TEZCAN/E+/Getty Images; cover, 3: Stefan Rotter/iStock/Getty Images; 4–5: (background) Andyworks/iStock/Getty Images; 5: (top) JNevitt/iStock/Getty Images, (bottom) lindsay_imagery/iStock/Getty Images; 6, 46–47: Subaqueosshutterbug/iStock/Getty Images; 7: Laurent Olivier/iStock/Getty Images; 8, 46–47: S.Rohrlach/iStock/Getty Images; 9: manuocen/iStock/Getty Images; 10: Gondwananet, via Wikimedia Commons (CC BY-SA 3.0); 11, 46–47: Lisa-ann Gershwin/ eAtlas.org.au, via Wikimedia Commons (CC BY 4.0); 12: Wing Ho/iStock/Getty Images; 13, 46–47: FtLaudGirl/iStock/Getty Images; 14: milehightraveler/E+/Getty Images; 15, 46–47: hayatikayhan/iStock/Getty Images; 16–17, 46–47: Mark Kostich/iStock/Getty Images; 18: Design Pics/Getty Images; 19, 46–47: twildlife/iStock/Getty Images; 20, 46–47: ePhotocorp/iStock/Getty Images; 21: Lensalot/iStock/Getty Images; 22: Dustin Rhoades/ iStock/Getty Images; 23, 46–47: TacioPhilip/iStock/Getty Images; 24: Oxford Scientific/ The Image Bank/Getty Images; 24–25, 46–47: through-my-lens/iStock/Getty Images; 26 (inset): Christina Williger/iStock/Getty Images; 26 (background), 46–47: marima-design/ iStock/Getty Images; 27: monicadoallo/iStock/Getty Images; 28, 46–47: kikkerdirk/ iStock/Getty Images; 29: helovi/iStock/Getty Images; 30 (top), 46–47: public domain, by Rodrigomorante, via Wikimedia Commons; 30 (bottom): public domain, by Centro de Informações Toxicológicas de Santa Catarina, via Wikimedia Commons; 31: Charles J. Sharp, via Wikimedia Commons (CC BY-SA 4.0); 32 (top, middle): scubaluna/iStock/Getty Images; 32 (bottom), 33, 46–47: LauraDin/iStock/Getty Images; 34–35: Coral_Brunner/ iStock/Getty Images; 35, 46–47: JaGr19/E+/Getty Images; 36, 46–47: Damiao Paz/iStock/ Getty Images; 37: heckepics/iStock/Getty Images; 38–39, 46–47: Queserasera99/iStock/ Getty Images; 40: camacho9999/iStock/Getty Images; 41, 46–47: Atelopus/iStock/Getty Images; 42: Anthony Paz – Photographer/iStock/Getty Images; 43, 46–47: ErikKarits/ iStock/Getty Images; 44, 44–45: Kwangmoozaa/iStock/Getty Images; 45, 46–47: frank600/ iStock/Getty Images; 46–47: (map) Fayethequeen/iStock/Getty Images; 48: agus fitriyanto/iStock/Getty Images

Visit us online at penguinrandomhouse.com.

Library of Congress Cataloging-in-Publication Data is available.

Manufactured in China

ISBN 9780593383964 (pbk)   10 9 8 7 6 5 4 3 2 1 WKT
ISBN 9780593383971 (hc)   10 9 8 7 6 5 4 3 2 1 WKT

PENGUIN YOUNG READERS

LEVEL
FLUENT
READER 4

TINY TERRORS!

THE WORLD'S SCARIEST SMALL CREATURES

by Ginjer L. Clarke

Some of the world's scariest animals are big—sharks, tigers, and bears. Oh my!

Many creatures
are tiny, more
deadly, and even
more terrifying!
Most of the animals
in this book are no bigger
than the size of your hands. But
they have venom, poison, stingers,
fangs, and much more. Are you
ready to meet these
tiny terrors?

Scary Swimmers

One of the most dangerous creatures
in the world is only six inches
long. It is so small that
you *could* hold it in
your hands. But
don't! The blue-
ringed octopus
has venom that
can kill you.

People swimming in warm oceans sometimes pick up or step on this little beauty. *Flash!* The octopus shines its blue rings as a warning. *Slash!* It bites when it senses danger. Its venom is incredibly strong. A person who gets bitten will likely stop breathing and may die.

The blue dragon sea slug is even smaller at only one inch long, or the length of a paper clip. It is deadly, too! This teeny creature floats upside down on the ocean waves. It blends in well with the water.

Man-of-war

Wham! The sea slug eats bits of a man-of-war. It stores the venom from this prey in its frilly fins. The venom becomes more powerful. *Bam!* The sea slug stings with the tips of its fins when a predator comes close.

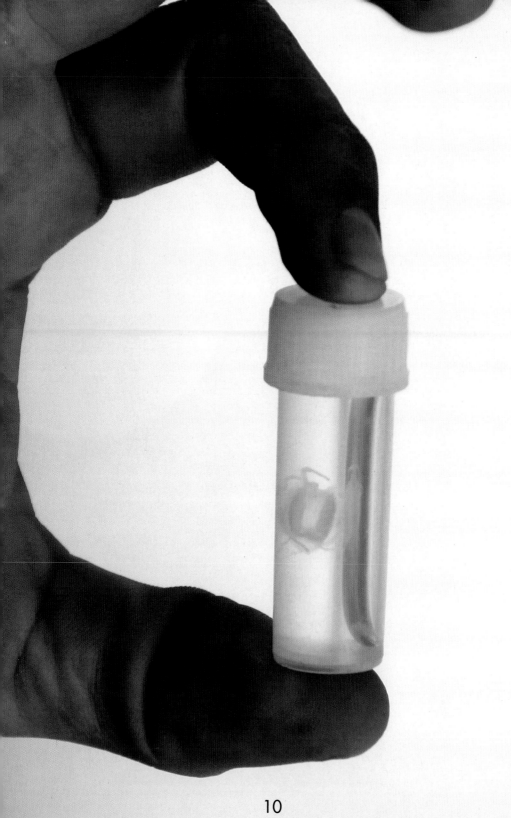

The Irukandji (say: IRR-oo-CAN-jee) jellyfish is the smallest jelly in the world. It is about the size of your pinkie finger. It stores venom in its tentacles. One Irukandji sting is 100 times stronger than a cobra's bite!

This mini-jelly is almost invisible in the ocean. *Zip!* It swims fast. *Whip!* Its tentacles can sting a swimmer. The jelly's venom causes extreme pain and sometimes death.

The puffer fish, also called a blowfish, does not *look* scary. A shark might even try to eat it. *Poof!* The puffer fish blows up its body into a spiky ball.

The shark gives up. If it did swallow the puffer fish, the shark would get a deadly surprise. The puffer fish has

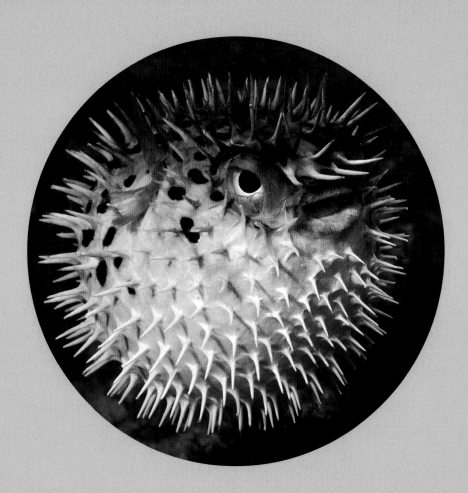

poison in its body that will kill anything that eats it. *Oof!*

Some people in Japan eat the poisonous puffer fish as a tasty treat. Chefs prepare the puffer fish carefully, but people can still die from eating cooked puffer fish.

Red-bellied piranhas are small, shiny fish. They gather in groups for safety. When it is time to eat, they attack together—hard and fast.

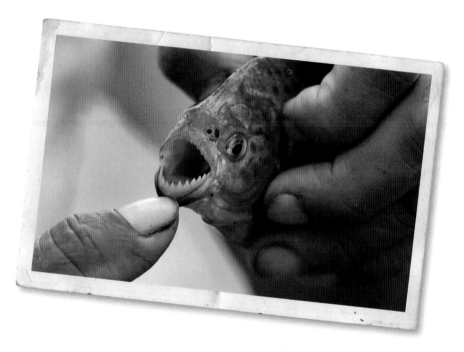

Munch! The piranhas bite the fins off a larger fish. Their sharp teeth rip and slice. *Crunch!* They can even eat a large dead mammal if it falls into the river. The hungry piranhas finish the huge meal in just a few minutes!

Creepy Creatures

The word *pygmy* (say: PIG-me) means very small. The pygmy rattlesnake is less than two feet long and weighs about one pound. But it is still dangerous!

Buzz! Buzz! The snake shakes the little rattle on its tail. It is warning a raccoon to back off. Its rattle is so quiet that the raccoon does not hear it.

The rattlesnake raises up its head and strikes quickly. It pokes the raccoon with its big front fangs. This puts lethal venom into its victim.

The Texas horned lizard gets its name from the horns, or spikes, on its head. These spikes help the teeny lizard blend into its desert home and protect it from enemies.

The lizard spots a hawk. It stays still and hopes the hawk does not see it. Uh-oh. The hawk *does* see it. The lizard

puffs out its body to look bigger. It jabs the hawk with its spikes. *Ow!*

The hawk keeps attacking. The lizard has one last amazing move. *Pow!* It squirts blood out of its eyes! The nasty smell scares off the hawk.

The Indian red scorpion is only two inches long. Yet it is the deadliest scorpion in the world! It sometimes sleeps in houses, shoes, and even beds. When a person touches the scorpion, it stings to defend itself. *Zap!* Its venom kills the victim if not treated quickly.

This itty-bitty scorpion usually only attacks its prey, not people. *Whap!* It stabs a lizard with its big red stinger. Then it grabs the prey with its claws.

The Brazilian wandering spider is
large for a spider at five inches wide. It
has the strongest venom of any spider!

This spider does not spin a web or hide
in a den. Instead, it wanders around the

forest hunting at night. It raises its front legs to show off its big red fangs. *Stab!* It kills a mouse with one bite.

It is also called the "banana spider" because it sleeps in banana plants. Sometimes one gets shipped to a grocery store by accident. *Jab!* The spider bites in self-defense if a person picks up the bananas.

The vampire bat gets its name because it drinks only one thing—blood! This small bat sleeps in a tree during the day. At night, it flies down to the ground. It lands near a cow and creeps closer.

Rip! The bat slices a circle of skin off the cow's leg with its sharp, pointy teeth.

Sip! It licks blood from the cut with its tongue for up to 30 minutes. The cow keeps sleeping. The bat's bite does not hurt, but it can give the animal rabies, a deadly disease.

Beautiful But Deadly

The fire-bellied toad has a secret. From the top it looks like a basic bumpy green frog. Underneath it hides a beautiful red belly. It also has a deadly surprise in its skin.

A snake comes close. *Flop!* The toad turns over and shows its bright colors. It tells the snake to keep away—or else. *Pop!* The toad oozes poison out of holes in its skin. The snake gets a taste and never bothers a fire-bellied toad again.

The golden poison dart frog is only one inch long. This itsy-bitsy beauty has the deadliest poison of any animal! Its skin holds enough poison to kill 10 humans. Don't worry, though. This frog does not hurt anyone unless they touch it.

Poison dart frogs get their name because some native people in the Amazon rain forest use the frogs' poison on blow darts. *Squoosh!* The people carefully scrape the poison from the frogs' skin. *Whoosh!* They shoot their darts into the trees to hunt animals.

A group of
caterpillars

The giant silkworm moth is harmless, but the caterpillar that turns into this moth is very dangerous. It has hundreds of small spines along its body. And they all contain venom!

This caterpillar lives on tree trunks. A person does not see it and accidentally touches it. The spines break off and stick into the person's skin. *Ouch!* The venom is very painful to humans and animals.

How can a snail be scary? When it has
a hidden spear-like tooth! Cone snails
come in many colors and patterns. Some
have fatal venom to use on their prey.

Fling! The cone snail shoots out its sharp tooth. A fish gets stung by the snail's venom and cannot move. The snail moves slowly toward its victim. *Sling!* Then it sucks the fish into its large mouth.

Look at those colors! The rainbow mantis shrimp is a beautiful creature. It is not a mantis *or* a shrimp, though—it is actually related to crabs and lobsters. It also eats them!

The rainbow mantis shrimp has big front claws. *Punch! Punch!* It jabs a crab and breaks the crab's shell. Its punch is faster than you can blink your eyes. It is also strong enough to break the glass of an aquarium. Whoa!

Fierce Fighters

Some other tiny, terrible creatures are bees. Not regular honeybees—those are helpful and important. These are killer bees!

Killer bees were created by scientists who wanted bees that could make more honey. Some of these bees accidentally

escaped into the wild many years ago.
Now they live in countries around
the world.

Killer bee stings are the same as
those of other bees. But killer bees are
more dangerous because they attack
in swarms. *Zoom! Boom!* Sometimes they
sting one person hundreds of times!

The Asian giant hornet has the frightening nickname of "murder hornet." This large insect has a long stinger that holds lots of venom. It also has big jaws—perfect for eating honeybees.

A swarm of giant hornets attack a honeybee hive. *Chomp!* They murder all the bees by biting off their heads. *Whomp!* They carry the bee bodies to their nest and feed them to their babies.

Bullet ants are not killers, but they have the most painful sting of any insect. They get their name because their sting is said to feel like being shot by a bullet. And the horrible pain can last for a whole day. Yikes!

These ants live in huge groups. They are only deadly when defending their nest. *Click! Click!* They make small sounds of warning before they attack. Bullet ants will sting some other ants many times to chase them away.

An assassin (say: uh-SASS-in) is
someone who kills an important person.
Assassin bugs kill other insects. And
there are many different types.

Some assassin bugs are also called
"kissing bugs." This is a nice name, but
they get it for bad behavior.

Kissing bugs bite people around the eyes or mouth while they sleep. Oh no! They also suck blood and spread disease. That is not nice!

What is the scariest animal of all? The mosquito.

Male mosquitoes drink nectar from plants. But female mosquitoes drink blood from mammals. A female mosquito zooms onto a victim. *Poke!* She jabs a human with her sharp mouthparts and sucks a little blood. But that does not hurt much.

So why are they the scariest? Mosquitoes can carry diseases in their saliva. Millions of people die every year from these diseases. See now why they are so scary? Be sure to wear bug spray when you are outside in the summer to protect yourself.

Most of these scary small creatures *do not* live in your backyard. *Phew!* That is good news.

This map shows where in the world
they can all be found.

Even the deadliest animals usually do not attack humans, except to defend themselves. So keep away from these tiny terrors and you should have nothing to fear!